W0082658

The
UNKILLABLES

JO LAMBELL

**40 RESILIENT HOUSEPLANTS
FOR NEW PLANT PARENTS**

JO LAMBELL

The UNKILLABLES

OH EDITIONS

PART 3

Plant SOS

Introduction

Are you a plant killer? I used to be. I bought three beautiful, exotic-looking houseplants and killed all of them with a toxic mix of overzealous watering and fussing. I knew I desperately wanted them to survive, but I had absolutely no idea what I was doing wrong. So, I did what every self-respecting adult would do and called my mum. With some basic knowledge and skills passed down to me, I felt confident enough to try again, and – success! – I managed to keep my first ever succulent alive.

Fast forward a few years and I'm a reformed character; now it's me who gets asked those very same questions about plants every day. It seems I wasn't ever alone in my plant-murdering ways.

If you've bought this book (or had it gifted to you) then you too have an interest in keeping your plant pals alive. It really is worth taking the time to understand why your houseplants keep crashing and burning (or wilting and dying). When given the chance to thrive, plants can bring so much joy to your life – they can transform your home into a gorgeous green space, give you something to nurture and care for and provide some amazing health benefits.

I'm here to take the fear out of the unknown and to equip those with no green fingers at all with the skills they need to keep their plants alive. Plants can be a little intimidating – with their fancy Latin names, complex terminology and 'just look at me and I'll perish' reputation – so choosing hardy, tolerant plants is the best place to start. (If you're not sure where to begin choosing a plant, try the Find Your Plant Soulmate quiz on page 12 and the Plant Cheat Sheet on page 14).

In this book, I'll walk you through the basics of care – light, water, location – and tell you when to repot your plants (spoiler: it's not winter) and what to do when things go wrong. I'll introduce you to many awesome varieties of unkillable plants. There are so many super-cool, high-maintenance-looking plants which not only tolerate neglect but actually thrive on it (I'm looking at you, Golden Pothos), so let's not waste time on the fiddly fusspots.

I promise you, by the time you finish this book you'll be a confident ex-plant-killer, just like me.

Getting Started

Find Your Plant Soulmate

Need a little help picking the perfect plant for you? Tot up your answers and all will be revealed.

How many plants do you own?	A	B	C	D
	A few	Lots	One	None – yet

Have you killed a plant before?	A	B	C	D
	My ratio is about 50:50	My plants are mostly thriving	Many, many times	This is my first attempt

How would you define your style?	A	B	C	D
	Maximal	Scandi	Boho	Minimal

Where do you want your plant to live?	A	B	C	D
	Office	Bedroom	Bathroom	Living room

Does size matter?	A	B	C	D
	The bigger the better	Little is lovely	I'd like a long, trailing plant	Somewhere in-between please

	A	**B**	**C**	**D**
Any preferences on leaf shape?	Statement and showy	Colourful and interesting	Flowy and not too uniform	Quirky and striking
How much care are you willing to give your new plant?	I'd quite like to fuss over my plant	I'm ready to commit	I'll try to remember to water it, but can make no promises!	I'm pretty busy and occasionally forgetful, so minimal
What superpowers would you like it to have?	Purify the air	Leaves open and close	Quick growing	Impossible to kill

MOSTLY As

A plant from the palm family would be perfect for you. You want a big, bold statement with maximum effect. Invite a Kentia Palm (page 44) into your home and enjoy instant good vibes. Bushy and beautiful, the Kentia will take centre stage in any room. It's also got a good chance of thriving where others may fail – in darker hallways or corners. It's air-purifying and resilient – the dream plant!

MOSTLY Bs

Looks like you should try a plant from the Calathea family. With bright, elegant leaves, these plants provide a bit of drama and a rainforest-vibe to any room. The classy Prayer Plant (page 38) will bring joy to any plant parent. Its gorgeous patterned leaves open and close at night (like hands closed in prayer). It's ideal for sprucing up a bedroom.

MOSTLY Cs

There's no judgement here, but you're probably a serial plant killer. That's ok! You're also quite impatient and want a plant that rewards you with some growth after you've shown it some (minimal) attention. The Golden Pothos (page 58) is here to get your plant parenting back on track. It's easy-going and quick-growing – potentially growing 30cm (12in) a month! It looks great trailing from a shelf.

MOSTLY Ds

If you're new to houseplants, you can't go wrong with a Snake Plant (page 41). This is one laidback plant, which looks at home in minimal interiors, providing texture and flowing lines. This is probably the toughest plant in the book. Near-impossible to kill and is as hardy as they come, it's the perfect way to welcome you to plant parenthood!

CHECK OUT THE PLANT CHEAT SHEET OVERLEAF TO FIND OTHER PLANTS IN THE BOOK THAT TICK ALL YOUR BOXES.

Plant Cheat Sheet

Plant		Max Size			Pet And Child Safe
		Up to 50 cm (20 in)	Up to 1 m (3 ft)	Over 1 m (3 ft)	
Prayer Plant	page 38	•			•
Snake Plant	page 40			•	
ZZ Plant	page 42			•	
Kentia Palm	page 44			•	•
Mikado Snake Plant	page 46		•		
Blue Star Fern	page 48		•		•
Yucca Plant	page 50			•	
Flaming Sword	page 52	•			•
Fishbone Cactus	page 54		•		•
Parlour Palm	page 56			•	•
Golden Pothos	page 58			•	
Bonsai Ficus	page 60		•		
Aloe Vera	page 62		•		
Spiral Cactus	page 64			•	•
Porcelain Flower	page 66		•		
Paper Plant	page 68			•	•
Cast Iron Plant	page 70		•		•
Hen and Chicks Succulent	page 72	•			•
Ivy	page 74		•		
Mistletoe Cactus	page 76		•		
Whale Fin	page 78		•		
Areca Palm	page 80			•	•
Sago Palm	page 82		•		
Sedum Burro's Tail	page 84		•		•
African Milk Tree	page 86			•	
Spider Plant	page 88		•		•
Bunny Ear Cactus	page 90		•		
Croton Plant	page 92		•		
Ponytail Palm	page 94			•	•
Brazilian Cactus	page 96			•	•
Steudner's Dragon Tree	page 98			•	
Golden Barrel Cactus	page 100	•			•
Dumb Cane	page 102			•	
Zebra Cactus	page 104	•			•
Peace Lily	page 106	•			
Jade Necklace	page 108	•			
Velvet Calathea	page 110		•		•
Heartleaf Philodendron	page 112			•	
Dragon Tree	page 114			•	
Birds Nest Fern	page 116		•		•

Air Purifying	Light Conditions		Watering Frequency			Maintenance Level
	Bright, indirect	Shade is fine	Semi-reg	Only when dry	Misting needed	0 = low ; 10 = high
•	•		•		•	5
•	•	•	•			1
•	•	•	•			3
•	•	•	•			4
•	•	•		•		2
•	•	•	•			6
•	•	•	•			4
•	•		•		•	6
	•			•	•	4
•	•	•	•		•	3
•	•		•		•	7
•	•		•			3
•	•			•		3
	•			•		2
•	•		•		•	6
•	•		•		•	6
•	•	•	•			2
	•			•		3
•	•		•		•	5
	•		•			4
•	•		•			5
•	•		•			7
•	•		•			7
	•		•			4
	•			•		5
•	•		•		•	4
	•		•	•		4
•	•		•		•	7
	•			•		5
	•		•			3
•	•	•	•		•	7
	•			•		2
•	•		•		•	5
	•			•		2
•	•		•			4
	•			•		1
•	•		•		•	8
•	•		•			5
•	•		•		•	7
•	•		•		•	8

Essential Kit

Before you get started on your journey to plant parenthood, there are some essential items you will need to get the best from your plants.

POTS

After buying your new plants, the second best thing is pot shopping! There are so many beautiful ways to house your greenery, from macramé hangers to giant woven baskets, chic metallic planters and everything in between. Remember always to look for decorative pots that will perfectly fit your plant and its nursery pot, to ensure good drainage. Don't be tempted to put your plant in a pot that is too big for it as it will struggle to absorb the nutrients from the soil. Aim for 1–2cm (¾in) space either side of your nursery pot.

COMPOST

It's a good idea to always have a bag of compost handy – you'll need it when it comes to repotting. A multipurpose compost will do the job just fine, but for a plant in the succulent or cacti families, opt for a specialist soil

MISTER AND WATERING CAN

Whilst it's tempting to use a glass or old jam jar to hydrate your new pals, you get better accuracy and control with a watering can. It's worth investing in a mister too – they're a brilliant way to increase humidity by mimicking raindrops, hydrating your plant right to the tips of its leaves – with most plants enjoying a weekly misting during warmer months.

SCISSORS

Like us humans, your plants will need the occasional haircut (page 26)! Invest in a pair of sharp gardening scissors to tidy up a leggy Porcelain Flower (page 66), or to remove crispy brown ends from a Kentia Palm (page 44). Look for scissors or snips with shorter blades for precision pruning and always keep them clean to avoid spreading disease. They'll also be your new best friend when you start propagating.

CLEANING CLOTH

Whilst the plants in this book are super hardy and require very little maintenance, big-leafed beauties will love you for wiping their leaves down with a cloth every couple of weeks. Dust particles settle on them, blocking their pores and making it difficult to photosynthesise, so gently wipe these off with a dry or damp cloth.

Plant Families

Let me introduce you to some of the most common plant families. Just like your own, no two plants in a family are the same, but there are some traits they all share. Plants from the same family have similar needs and like similar conditions, so once you've found a spot one likes, you can add some of its siblings!

SUCCULENTS

Known for their beautiful, fleshy leaves, succulents are immensely popular for many reasons, but mainly because they're so hard to kill, making them the perfect starter plant. These clever little plants store water in their leaves, and use it when they need it, so require very little watering. They love the sun and will thrive on a sunny windowsill or shelf. No self-respecting collection should be without at least one or two.

Try: Hen and Chicks Succulent (page 72), Sedum Burro's Tail (page 84), Zebra Cactus (page 104), Jade Necklace (page 108)

CALATHEAS

Famed for their bright colouring and interesting leaves, calatheas win first prize in the looks department. They do well in indirect light and enjoy humid conditions, so make brilliant bathroom plants. Some members of this family can be a little tricksy, but don't tarnish them all with the same brush – only the hardiest varieties are included in this book to get you started. Weekly misting is the key to caring for these varieties – you heard it here first!

Try: Prayer Plant (page 38), Velvet Calathea (page 110)

PALMS

If you're looking for maximum impact and minimum maintenance, a palm is the perfect plant for you. With their lush, green, feathery foliage, they instantly transform any room. They're really easy to care for, which adds to their appeal, thriving where other plants may fail, in shady spots like hallways or rooms with low light levels.

Try: Kentia Palm (page 44), Parlour Palm (page 56), Areca Palm (page 80)

FERNS

Ferns have a little bit of a reputation for being fussy – even the most experienced indoor gardeners balk at some varieties. But not all ferns are made the same, and I've chosen the toughest cookies for you to get to know. With interesting and exotic leaves, a fern is a great addition to your green gang – they'll be right at home in your kitchen or bathroom where there are higher levels of humidity.

Try: Blue Star Fern (page 48), Birds Nest Fern (page 116)

CACTUS

The cactus is the ultimate unkillable family. From the small, cute and spiky to the long, trailing and spine-free, there are so many interesting varieties to choose from. And, of course, they are really easy to look after – super adaptable, you can place them in virtually any room and they'll thrive.

Try: Fishbone Cactus (page 54), Spiral Cactus (page 64), Mistletoe Cactus (page 76), African Milk Tree (page 84), Bunny Ear Cactus (page 90), Golden Barrel Cactus (page 100)

Plant Locations

Finding the perfect home for your plants is easy – just consider the amount of light, heat and humidity your plant needs to thrive and start creating your own green oasis.

KITCHEN

Everyone knows, the real hub of the house is the kitchen – it's the engine room that fuels the family and is as worthy of greening up as any other space in your home. Pop some succulents (page 18) on open shelves to elevate the space and add some interest to a breakfast bar with a Bonsai Ficus (page 60). A Cast Iron Plant (page 70) in a gorgeous planter also looks fab alongside your cabinets and Ferns will thrive in a steamy space. When placing your plants, be wary of bright afternoon sun, which can scorch leaves.

LIVING ROOM

One of the easiest rooms to fill with plants is the living room, as most plants thrive well here (provided there is adequate natural light). The ideal temperature for houseplants is 15–24 degrees Celcius (59–75 degrees Fahrenheit), and you'll find living rooms have the most stable temperature. It's a real opportunity to surround yourself with greenery – pop a giant Kentia Palm (page 44) in the corner of the room, trail a Porcelain Flower (page 66) from bookshelves, line your windowsill with Hen and Chicks Succulents (page 72) or pop a Peace Lily (page 106) on your coffee table – once you start, you wont be able to stop!

OFFICE

Whether your desk is at home or in an office, you can benefit hugely from some green work fellows. A Bunny Ear Cactus (page 90) is a great desk buddy – you focus on work, and they won't mind if you forget about them for a while. Snake Plants (page 40) are brilliant at zapping free radicals and can handle low light if you work somewhere with little natural light. Or add a Prayer Plant (page 38) – a lovely way to bring some zen to your desk as you watch its leaves move up and down throughout the day.

BEDROOM

Struggling to drift off? Instead of counting sheep, consider adding some houseplants to your bedroom. Greenery can help lower your heart rate and reduce the stress hormone cortisol. Not to mention, they're air-purifying so can help you to breathe better. Try an Aloe Vera (page 62), a Spider Plant (page 88) or a Birds Nest Fern (page 116) – all of which rank highly for their air-purifying benefits.

`BATHROOM

Your bathroom is the greenhouse you didn't know you had – often with minimal natural light and higher humidity levels, it takes a special kind of plant to thrive here. Try draping a Golden Pothos (page 58) from a shelf, curtain rod or windowsill to maximise space or add a beautiful Dragon Tree (page 114) by the window. Peace Lilies (page 106) love to have moist soil, so the steam from this room suits them well.

Seasonal Gardening

Even though the plants featured in this book are 'unkillable', it is still important to make sure you're giving them the right care all year round with these simple seasonal tips.

In Spring

INCREASE WATERING

Some plants go through a dormant period in winter, where they require less water. Naturally, the arrival of spring signals the end of this and plants enter growing season so require hydrating more frequently.

CUT BACK DEAD LEAVES

Use a sharp pair of scissors to remove any browning or yellow leaves – this will encourage fresh new growth and denser, more luscious foliage will grow back. Try not to confuse new growth with older yellowing leaves – new leaves often come through yellow or lime in colour, darkening as they mature.

REPOT AND REFRESH

Motivate your plant to bloom and grow by replenishing its soil with nutrient-rich compost. Now is the time to consider repotting into a bigger container, too (page 24). Simply tap your nursery pot away from the soil – if the roots are densely packed or bursting through the holes in the pot, it's time.

In Summer

ROTATE PLANTS REGULARLY

The sun is out! Ensure its rays are distributed evenly among your plants by turning them weekly, so all sides of the plant get equal exposure. Twisting the pot a quarter turn each week will ensure your plant grows full and even, not lop-sided or top heavy.

TAKE THEM OUTSIDE

Treat (some of) your greenery to a little holiday by letting them sit outside during summer. Spider Plants (page 88), Yuccas (page 50) and many cacti all enjoy soaking up the sunlight. Just remember to bring them inside when the temperature drops.

TOP UP THEIR HUMIDITY

Plants that like high humidity (especially the tropical types like Calatheas and Ferns) should be misted frequently through periods of summer heat – spray under leaves too so the water falls on to the lower, less visible leaves.

In Autumn

EASE UP ON WATERING

Your plants won't need a drink as regularly during the cooler months as they won't be using as much energy to grow. Allow their soil to dry out a little more between watering during this time. It's not uncommon to only water some cacti as little as once a month.

AVOID DRY AIR

As the temperature drops, the radiators and central heating go on! Move your plant away from any heating vents to avoid upsetting them with the dry air – at least 1 m (3 ft) from the heat source.

INCREASE LIGHT

The shorter days means less natural daylight for your plants. South-facing windows tend to see more sunlight, so consider moving them here, or alternatively you could buy a grow light to help top up its much-needed rays.

In Winter

DON'T FERTILISE

Now is the time to ditch the fertiliser and let your plants slide into winter slumber. Repotting and propagating are jobs for spring, and now is the time for your greenery to rest.

DUST THEM

In a season where daylight is already scarce, having dusty leaves decreases the amount of light that can reach the surface of your plant's leaves, making it difficult for them to make food. Gently remove dirt with a damp cloth.

KEEP THEM WARM

If temperatures go below 10°C (50°F) you may see your plants suffer. Aim to keep them homed somewhere with a consistent temperature that doesn't fluctuate. Move them away from draughty windows – the cold can send your plants into shock.

Repotting

Repotting is an essential part of your houseplant's care routine, and it's not as difficult as you may think! Typically, your green gang will need to be repotted every 12–18 months, depending on how quickly they grow. If you notice roots coming through the bottom of your plant's current container or if it's top heavy and falls over easily, these are clear signs that your plant is ready for a bigger home. Spring marks the beginning of growing season which is the perfect time to repot. Avoid repotting when your plant is in a dormant period, such as winter, as its pace of growth will have slowed.

1. Size up sensibly

Choose a new pot that is around 3–5 cm (1¼–2 in) larger in diameter than your plant's current container. Try not to go beyond two pot sizes bigger – anything larger than this can overwhelm the plant with soil and water it doesn't need.

2. Remove carefully

Ease your plant out of its current pot by turning it upside down or at a slight angle. Rotate the plant to loosen it and tap on the bottom until it slides out.

3. Reorganise the roots

Before repotting your plant, take a closer look as its roots. Prune any older roots and carefully loosen and untangle them with your fingers so they grow outwards. By doing this, your plant will know it has extra room to spread out in its new container.

4. Freshen up the soil

Repotting isn't only about a bigger pot, it's also a chance to pack in a layer of new soil, and consequently fresh nutrients, for your plant to enjoy. Choose the correct soil for your plant – a multipurpose peat-free soil is great for the majority of your plants and choose a cactus potting mix for your spiky friends. Set the plant in the new soil, ensuring it can stand upright on its own. If your new pot doesn't have a drainage hole, add a layer of pebbles to the bottom layer to help prevent waterlogging.

5. Water thoroughly

To finish the process of repotting, water your plant generously. It's essential that any excess water can drain to the bottom of the pot and escape once your plant has been hydrated.

Pruning

Houseplants have more in common with us humans than you may realise. Just like us, they like to look and feel their best, which is why pruning is an essential part of their care routine – think of it like a haircut for plants. Trimming away any dead or overgrown stems, leaves and vines will encourage fresh and healthy growth. Pruning also allows you to maintain your plant's shape and size, keeping it suitable for indoor spaces. In other words, it's a great way to prevent your greenery from taking over your home!

Typically, you should aim to do this at the start of growing season, which is early spring, so your plant can flourish into the warmer months with fresh and strong leaves. Pruning is important because damaged leaves and stems can be an energy drain to the plant which will slow its growth. If some of the leaves are damaged due to disease, removing them can stop this spreading to the rest of the plant.

All you'll need to give your plant a good trim is a pair of clean and sharp scissors. Blunt or dull blades can result in a sloppy cut, and dirty scissors can easily transfer disease or pests. Start by removing any dead or browning leaves. Just don't get too scissor-happy – as a rough guide, cut away around a quarter of your plant's existing leaves.

If your plant has leggy stems, prune these right back to a node (where the leaf meets the stem). You'll be surprised by how quickly new growth appears! As for trimming vines, cut directly below a leaf as this will keep your plant looking compact and full. Some plants, like Golden Pothos (page 58) grow quite aggressively towards the light source which means they can become too big for their locations – pruning helps to control this.

While the majority of plants enjoy a little pruning, there are some that don't require this kind of maintenance – succulents and snake plants need very little trimming, and will actually suffer more from an unnecessary haircut, so only remove their dead leaves.

Propagating

Propagation is a rewarding process involving taking cuttings from your existing plant to produce plant babies. In other words, you're creating new plants, for free! You don't need to be an expert to expand your green family either.

While there are many methods of propagation, cutting from the stem works best on a plant that has vine-like stalks, such as the Golden Pothos (page 58) and Heartleaf Philodendron (page 112). Spider Plants (page 88) and Fishbone Cacti (page 54) are also super easy to propagate.

1. Choose your stem

Identify a healthy stem that you wish to cut from and remove with a sharp pair of scissors to make sure the stem comes away cleanly, without damaging your plant. Choose a stem that is thriving and has plenty of new growth or leaves on it. This will ensure you are reproducing from the healthiest parts of your plant.

2. Place in water

Next, fill a small vase or jam jar with water and insert the cutting into this, stem first. And so begins the waiting game! In the meantime, leave this container somewhere with lots of bright indirect light.

3. Keep an eye on it

Refresh the water when necessary as it may evaporate. Aim to do this every week or so to keep conditions fresh. After as little as 2–3 weeks, you should start to see tiny roots appear. Ideally, these roots need to be at least 2.5 cm (1 in) in length before they are moved out the water.

4. Pot your cutting

Now it's time to plant the cutting. Use a small pot (around 9 cm/3.5 in) so it doesn't get overwhelmed, and fill with a multipurpose houseplant compost. Use your finger or a pencil to insert a hole into the soil. Place your new root into the hole and push the soil around it firmly. Water until it comes through the holes at the base of the pot and sit on a saucer for 30 minutes until the water has had a chance to drain through and settle. Position your new plant somewhere dry with lots of bright indirect light and carefully monitor its progress. The cutting should eventually grow into a thriving plant.

Cleaning

You wouldn't buy a brand-new car and then never take it to the car wash – and the same applies to your plant! Remember to keep them looking and feeling great with these easy tips to freshen them up.

DONE AND DUSTED

The waxy texture of some leafy plants can be like a magnet to dust. Dust is actually very damaging to plants as it blocks their stomata (plant pores), preventing successful photosynthesis meaning your plant can't grow. Gently remove this with a damp cloth, and avoid using chemicals or shine spray which can be damaging.

IN THE MOOD FOR FOOD

During growing season, give your plants a helping hand by providing them with some extra nutrients in the form of plant food. There are many solutions that you can mix into water when giving them a drink. Aim to do this every other watering, and only do this during growing months.

REORGANISE AND REGROUP

If you assemble your botanical beauties together as a group, consider swapping around their usual formation so that they're getting an equal distribution of light. Do this every couple of months, as most plants don't like moving about too much – they'll let you know if they're unhappy by dropping their leaves.

Going on Holiday

Whether you're driving home for Christmas or jetting off somewhere exotic, chances are at some point you'll have to leave your houseplants unattended. Don't worry, there are some things you can do to keep them happy while the cat's away.

1. Give them a drink

You'll want to make sure they stay hydrated whilst you're away so give them a good watering before you go. If it's a winter break, plants tend not to require as much watering anyway, so you could try the ice cube hack (page 122).

2. Put them in your bath

This might sound ridiculous, but hear us out! If you have plants that need lots of humidity, then consider soaking some old towels and newspapers in water, placing them in the bath, then put your plants on top. Et voilà – moisture and humidity for your green friends.

3. Ask a neighbour to call in

Everybody needs good neighbours, including your plants! If your plant family requires a little more care, ask a neighbour you trust to pop round and tend to their needs. You'd do the same for your pet, after all.

4. Give them a haircut

Before you leave, give your plants a quick spruce up – cut away any dead leaves (page 22) as this encourages them to grow new ones and keeps them looking and feeling healthy. Remove any fallen debris from the soil and dust them down.

5. Group them together

Your plants are stronger as a pack and grouping them together helps them to thrive. Move some of your tropical plants to within close proximity of each other to create their own microclimate – this will increase humidity levels, giving them the moisture they crave.

6. Try a pebble tray

A great way to increase moisture and eek out some time between waterings is by using a pebble tray. Line a saucer with small stones or pebbles. Add water, but make sure the pebbles are still visible above the surface – this means your plant won't be sitting in water, but will still have access to moisture.

PART 2

The Plants

Prayer Plant

MARANTA LEUCONEURA

Ah, the beautiful Prayer Plant. It's one of the most popular houseplants around, and with good reason, too – just look at its handsome leaves! It acquired its name due to the way its leaves lie flat during the day and fold together at night, like hands closed in prayer – cool, hey? Calathea plants (the family the Prayer Plant belongs to) have a bit of a reputation for being divas; however, the Prayer Plant is surprisingly low maintenance.

This plant is native to the rainforests of Brazil, so replicate the conditions of its natural habitat by misting it regularly. This should also stop the tips of the leaves going brown or crispy – a common Prayer Plant problem. It will be quite happy in a well-lit bathroom, enjoying the steam from the shower.

Water around once a week, keeping the soil moist but not soggy, and reduce this frequency during winter months. Make sure your Prayer Plant has plenty of bright, indirect light.

HOW TO KEEP ME HAPPY

Light ☼
I like bright indirect light

Water ◊
Water me when my top 5 cm (2 in) of soil feel dry to the touch

Air-purifying ≋
I remove the nasties lurking in your room

Safety ⚠
I'm safe around your furry friends and little people

Size ↗
Up to 30 cm (12 in)

JO'S TOP TIP

This plant can be a little sensitive to the chemicals found in tap water; if possible, use rain water which has a lower pH.

Snake Plant/ 'Mother-in-Law's Tongue'

SANSEVIERIA LAURENTII

While there are many hardy plants in this book, the Snake Plant takes the crown for being the toughest kid on the block. Don't let the name fool you, though – this plant is incredibly laid back and is perfect for beginners.

The Snake Plant is a real show-stopper with its long variegated leaves, and provides texture and colour to any room that homes it. It doesn't need too much water or light to survive, but its leaves do have a waxy texture and can be prone to dust, which hinders photosynthesis, so they will appreciate a wipe down when this happens.

The Snake Plant prefers to be on the drier side and can be susceptible to root rot, so avoid overwatering and make sure its pot has sufficient drainage. Never let your Snake Plant sit in water and you'll be rewarded with a happy, healthy houseplant.

HOW TO KEEP ME HAPPY

Light ☼
I like bright indirect light, but can tolerate shadier spots

Air-purifying ≋
I remove the nasties lurking in your room

Water ◊
Water me when my top 5 cm (2 in) of soil feel dry to the touch

Safety ⚠
Keep me out of reach of furry friends and little people

Size ↗
Up to 1.2 m (4 ft)

JO'S TOP TIP
Keep this plant in your bedroom or office as it will store oxygen throughout the day and then release it back into the atmosphere at night.

ZZ Plant

ZAMIOCULCAS ZAMIIFOLIA

Affectionately named the ZZ Plant (a much-easier-to-say abbreviation of its Latin name), this tropical plant is as hardy as they come. Native to Zanzibar, Kenya and eastern Africa where it enjoys a warm climate, the ZZ has the ability to store water – great news for plant newbies or those that lead a busy life, as it'll survive should you forget to water it.

The ZZ is very well loved and for good reason. Its glossy leaves reflect light, brightening any room, and its care instructions are almost non-existent, too. Low light? It will handle it. Low water? No problem. Its only real ask is a regular wipe down to keep dust from blocking the pores of those shiny leaves.

Does the mighty ZZ have any downfalls? Not really, but it can be prone to root rot so avoid killing with kindness by overwatering. If your ZZ has dark mottling on its leaves, it's a sign of too much sun, too much water or infestation so pay attention if you see this symptom.

HOW TO KEEP ME HAPPY

Light ☼
I like bright indirect light, but can tolerate shadier spots

Water ◊
Water me when my top 5 cm (2 in) of soil feel dry to the touch

Air-purifying ≫
I remove the nasties lurking in your room

Safety ⚠
Keep me out of reach of furry friends and little people

Size ↗
Up to 1.2 m (4 ft)

JO'S TOP TIP
If your ZZ starts to spread and lean, use twine to tie the stalks closer together.

Kentia Palm

HOWEA FORSTERIANA

The Kentia Palm is the plant world's BFG (big friendly giant), and its epic leaves are real crowd-pleasers. In fact, Queen Victoria loved these plants so much, she had them in all of her homes. If it's good enough for royalty... With its feathery green fronds you'd be forgiven for thinking it would be a tricky customer, but the Kentia is highly adaptable.

This is a slow-growing plant, with an impressive wing span, so make sure it has the space it deserves. Able to deal with the lowest levels of light, it makes the perfect showstopper for a darker room or hallway. Mist it regularly, and you will be rewarded with growth.

Run a damp cloth over its leaves to keep the Kentia Palm dust-free. It will reward you with its incredible air-purifying skills, removing toxins such as formaldehyde and ammonia often found in cleaning products from your home.

HOW TO KEEP ME HAPPY

Light ☼
I like bright indirect light, but can tolerate shadier spots

Safety ⚠
I'm safe around your furry friends and little people

Water ⬦
Water me when my top 5 cm (2 in) of soil feel dry to the touch

Size ↗
Up to 2 m (6½ ft)

Air-purifying ≋
I remove the nasties lurking in your room

JO'S TOP TIP
The Kentia Palm's leaves will turn brown if the air is too dry or if it needs more water. If you are watering it *too* frequently, yellow tips may appear on the leaves.

Mikado Snake Plant

SANSEVIERIA BACULARIS MIKADO

Yes, this plant does look familiar – it's a compact hybrid of the classic Snake Plant. Named after its tall, spikey fronds, Bacularis comes from the Latin word 'baculum' meaning rod or stick. In true Snake Plant style, it's super hardy and very low-maintenance and a great air purifer, too.

This plant is sometimes referred to as Witches' Fingers on account of its appearance – I think it's far more elegant though!

This plant is native to West Africa, so it is accustomed to dry conditions. Be careful not to overwater it – the Mikado prefers its soil to dry out between watering. It also fairs well in shady spots, so you can place it wherever you like!

HOW TO KEEP ME HAPPY

Light ☼
I like bright indirect light

Water ◊
Let me dry out between watering

Air-purifying ≋
I remove the nasties lurking in your room

Safety ⚠
Keep me out of reach of furry friends and little people

Size ↗
Up to 1 m (3 ft)

JO'S TOP TIP

This plant stores water in its leaves – give them a gentle squeeze and if they yield very easily, it means your plant is thirsty.

Blue Star Fern

PHLEBODIUM AUREUM

The Blue Star Fern is a quirky character, with its elongated green-blue fronds and fuzzy brown rhizome. This plant is an epiphyte, which means that in the wild, it grows on other plants and trees, rather than directly in the soil. In its natural habitat, its leaves have been known to reach 1.3 m (4 ft) in length – don't worry, it won't get that big in your home.

Keep your fern somewhere well-lit but out of direct sunlight to encourage growth. You can even place it outside during warmer months, but remember to bring it back indoors when winter approaches.

The Blue Star Fern enjoys a humid environment much like its native home, the tropical rainforests of South America, so mist it regularly and ensure the soil is moist, not soggy. It will be happiest in the bathroom or kitchen. Water weekly during the warmer months or when the soil appears dry, and less so throughout winter.

HOW TO KEEP ME HAPPY

Light ☼
I like bright indirect light, but can tolerate shadier spots

Water ◊
Water me when my top 5 cm (2 in) of soil feel dry to the touch

Air-purifying ≋
I remove the nasties lurking in your room

Safety ⚠
I'm safe around your furry friends and little people

Sizing ↗
Up to 1 m (3 ft)

JO'S TOP TIP
Blue Star Ferns don't appreciate water being poured directly into the heart of the plant, so water from the sides.

Yucca Plant

YUCCA ELEPHANTIPES

The Yucca Plant delivers on all fronts – it's sturdy, resilient and striking. Native to Mexico, the Yucca has distinctive sword-like leaves and grows strong, with no need for stakes or support. Although this is a slow-growing plant, stick with it, as it can eventually reach a height of 3 m (10 ft) indoors.

There are about 40 species in the Yucca family. This Yucca is one of the few plants that can cope with direct sunlight, which helps it to grow taller. Try not to overwater it – if the roots are soggy, hold off before watering again and ensure there is adequate drainage.

This plant can withstand a little unintentional neglect; however, its leaves need an occasional dusting with a damp cloth to unblock its pores so it can photosynthesise properly.

HOW TO KEEP ME HAPPY

Light ☼
I like bright indirect light, but can tolerate shadier spots

Safety ⚠
Keep me out of reach of furry friends and little people

Water ◊
Water me when my top 5 cm (2 in) of soil feel dry to the touch

Size ↗
Up to 2.5 m (8 ft)

Air-purifying ≋
I remove the nasties lurking in your room

JO'S TOP TIP
Yucca plants are surprisingly easy to propagate. Turn to page 28 to find out how.

Flaming Sword

VRIESEA 'ASTRID'

The Flaming Sword is a powerful name and it's certainly fitting for this houseplant. One of the more common bromeliads, it flowers for 3–6 months of the year, and its blooms are spectacular. Don't be put off by its high-maintenance appearance, it's actually very easy to care for.

There are 250 varieties of Vriesea and this plant has origins in South and Central America. It's an epiphyte in its native habitat, which means that despite its tendency to grow on other plants (rather than in the soil), it doesn't harm them in the process.

This is a tropical plant, so you'd be forgiven for thinking it is a lot of work to look after. It will appreciate humidity and misting; try grouping your tropical plants together to naturally increase humidity.

HOW TO KEEP ME HAPPY

Light ☼
I like bright indirect light

Water ◊
Water me when my top 5 cm (2 in) of soil feel dry to the touch

Air-purifying ≫
I remove the nasties lurking in your room

Safety ⚠
I'm safe around your furry friends and little people

Size ↗
Up to 50 cm (20 in)

JO'S TOP TIP

The Flaming Sword plant will do well near a window that's east- or west-facing; just be careful to avoid direct sunlight as this can burn its beautiful leaves.

Fishbone Cactus

EPIPHYLLUM ANGULIGER

Nothing fishy here! This unusual and intriguing cactus originates from the rainforests of Mexico and has uniquely shaped long, flat stems resembling a fishbone, hence its name. Its stems begin by shooting upright, but as they grow longer they'll begin to trail down.

Unlike a regular arid-loving cactus, keep your Fishbone happy with regular misting and moist soil – it loves humidity and bright indirect light. Don't be surprised to see its fronds standing upright – a sign it is hydrated right to its tips!

Its super-long stems can be pruned back if it's looking a bit leggy, and new stems will usually grow from the cut, making the plant look fuller. Its zig-zag stems look great hanging or trailing from a shelf.

HOW TO KEEP ME HAPPY

Light ☼
I like bright indirect light

Water ◌
Let me dry out
between watering

Wellbeing ♡
I can help you to relax and
boost your mood

Safety ⚠
I'm safe around your furry
friends and little people

Size ↗
Up to 1 m (3 ft)

JO'S TOP TIP

Feeling confident? This fishbone is a great plant to propagate – simply cut a stem and place in water until roots appear, then plant in soil. You can do it!

Parlour Palm

CHAMAEDOREA ELEGANS

This tropical plant is still as popular now as it was almost 150 years ago – the Victorians loved it and so will you! Known for its long stems and lush green feathery foliage, it not only looks fantastic but is a great air purifier. The Parlour Palm is native to the rainforests of southern Mexico, so has the ability to adapt to relatively low light levels and low temperatures.

When grown inside, the Parlour Palm can reach up to 1.2 m (4 ft) in height, but it is slow growing, so don't expect this growth overnight.

The Parlour Palm will appreciate humid conditions and the occasional misting. If you notice the leaf tips turning brown, this means the humidity levels aren't high enough. It can also cope with some unintentional neglect, not that I'm advocating this at all!

HOW TO KEEP ME HAPPY

Light ☼
I like bright indirect light, but can tolerate shadier spots

Safety ⚠
I'm safe around your furry friends and little people

Water ◊
Water me when my top 5 cm (2 in) of soil feel dry to the touch

Size ↗
Up to 1.2 m (4 ft)

Air-purifying ≫
I remove the nasties lurking in your room

JO'S TOP TIP

This is a slow-growing plant, but you can help it flourish by adding fertiliser to its soil during growing season.

Golden Pothos/ Devil's Ivy

EPIPREMNUM AUREUM

The Golden Pothos is a super-hardy houseplant that is perfect for new plant parents. It will grow long vines, so looks great on shelves or as a hanging plant.

This plant is sometimes referred to as Devil's Ivy as a reference to how it grows in the wild up the trunks of trees and often in shade. It's virtually impossible to kill, which is great news for any serial plant killers. If you've forgotten to water it, don't worry – those sad leaves will perk right up again with a drink.

Ensure this plant is kept moist with weekly watering and regular misting, although you should reduce the frequency of watering during the winter months. It's a fast-growing plant, so feel free to give it a trim if the vines get too long.

HOW TO KEEP ME HAPPY

Light ☼
I like bright indirect light

Water ◊
Water me when my top 5 cm (2 in) of soil feel dry to the touch

Air-purifying ≋
I remove the nasties lurking in your room

Safety ⚠
Keep me out of reach of furry friends and little people

Size ↗
Up to 2.5 m (8 ft)

JO'S TOP TIP
This is one of the easiest plants to propagate. Turn to page 28 to find out how.

Bonsai Ficus

FICUS GINSENG

Bonsai tree but for beginners? Enter the Ficus Ginseng. With large roots and oval leaves, this is a member of the Mulberry family and is found across southern Asia. There are more than 2,000 varieties of Ficus trees worldwide and this one isn't just a pretty face – ginseng has been used in Chinese medicine for centuries to cure all sorts of ailments, from stress and fatigue to sore throats, although I'm perfectly happy with it just looking cute in my home.

If it's possible for a plant to be considered old and wise, then the Ficus Ginseng definitely is. The roots are often grown for around 15 years in specialist nurseries.

The Ficus Ginseng is a simple plant, requiring a minimal care routine of warmth and indirect sunlight. Keep it out of draughts and windy spots as they hate the cold, and you're good to go!

HOW TO KEEP ME HAPPY

Light ☼
I like bright, indirect light

Water 💧
Water me when my top 5 cm (2 in) of soil feel dry to the touch

Air-purifying ≋
I remove the nasties lurking in your room

Safety ⚠
Keep me out of reach of furry friends and little people

Size ↗
Up to 60 cm (2 ft)

JO'S TOP TIP

Who says plants can't go on holiday? In the warmer months, it may like to sit outside on your patio (provided it's not in direct sunlight), just mind the temperature doesn't drop below 12–15°C (53–59°F).

Aloe Vera

ALOE BARBADENSIS

I love a multi-tasker, and they don't come better than the Aloe Vera plant. Not only hardy and easy to grow, it also boasts some awesome healing powers. Inside its stems is a clear gel containing 75 health-giving nutrients, perfect for soothing insect bites, burns and sunburn.

In the wild, the Aloe Vera can grow to 1 m (3 ft) in width. Don't worry about that happening in your home; however, it can become very heavy, so keep it in a sturdy pot.

The Aloe Vera is easy to care for and tolerant of the occasional missed watering. Pruning is generally not necessary either. Be careful not to over-water this plant because if the soil is wet day after day, its roots could start to rot.

HOW TO KEEP ME HAPPY

Light ☼
I like bright indirect light

Water ◊
Let me dry out between watering

Air-purifying ≋
I remove the nasties lurking in your room

Safety ⚠
Keep me out of reach of furry friends and little people

Size ↗
Up to 60 cm (2 ft)

JO'S TOP TIP

Not enough light can cause this plant to become dormant and the leaves have been known to droop downwards. If you have a sunny windowsill, pop it there – the hardy Aloe is one of the few plants that can withstand direct sunlight.

Spiral Cactus

CEREUS FORBESII SPIRALIS

Native to South America, the Spiral Cactus is a real statement houseplant and very low maintenance, too. Easily recognisable for its tall and twisted blue-green form, each cactus is perfectly unique. It begins to grow into a twisted shape all by itself once it has reached a height of around 10 cm (4 in) – it doesn't need to be trained to do so.

The Spiral Cactus can eventually reach over 3 m (10 ft) in height, so pop it somewhere with some room to stretch upwards! With the right care, you may even be lucky enough to see it flower during the summer months.

This is a very easy plant to care for, perfect for those with busy lives or who are new to plant parenthood – the only golden rule is simply not to overwater it. It is drought tolerant, so best to water only when the soil starts to dry out.

HOW TO KEEP ME HAPPY

Light ☼
I like bright indirect light

Water ◊
Let me dry out between watering

Wellbeing ♡
I can help you to relax and boost your mood

Safety ⚠
I'm safe around your furry friends and little people

Size ↗
Up to 3 m (10 ft)

JO'S TOP TIP

Is your cactus getting dusty? Use an old paintbrush or make-up brush to gently remove any cobwebs or dust.

Porcelain Flower

HOYA LINEARIS

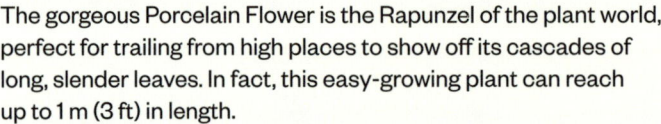

The gorgeous Porcelain Flower is the Rapunzel of the plant world, perfect for trailing from high places to show off its cascades of long, slender leaves. In fact, this easy-growing plant can reach up to 1 m (3 ft) in length.

With the right exposure to light, this plant will reward you with clusters of cream flowers during spring and summer, which some say have a lemony scent to them.

The Porcelain Flower is easy to care for and can be pruned if you wish to control the length and fullness of its stems to stop it becoming leggy. It will appreciate regular misting, aiming for moist not soggy soil, but remember to reduce watering during winter when it becomes dormant.

HOW TO KEEP ME HAPPY

Light ☼
I like bright indirect light

Water ⬡
Water me when my top 5 cm (2 in) of soil feel dry to the touch

Air-purifying ≋
I remove the nasties lurking in your room

Safety ⚠
Keep me out of reach of furry friends and little people

Size ↗
Up to 1 m (3 ft)

JO'S TOP TIP
If the leaves start to look a little shrivelled, this is usually a sign of thirst.

Paper Plant

FATSIA JAPONICA

The Paper Plant is native to Japan, southern Korea and Taiwan (China), where it enjoys humid conditions and dislikes the cold and wind. The Paper Plant has eight leaf lobes, which is where it gets its Latin name from: 'fatsia' is the Japanese word meaning 'eight'. There are both indoor and outdoor varieties.

Keep yours happy with regular misting, but be careful of overwatering as this plant is prone to root rot. It also likes to spread its leaves out wide, so give your Paper Plant the space it needs to grow.

You'd be forgiven for thinking this tropical-looking plant is high-maintenance, but in reality, the Paper Plant is as easy-going as they come and can grow to an impressive size without much effort. It will handle most conditions, but to see it thrive, give it an equal mix of sun and shade and avoid draughts and direct sunlight.

HOW TO KEEP ME HAPPY

Light ☼
I like light bright indirect light

Water ◊
Water me when my top 5 cm (2 in) of soil feel dry to the touch

Air-purifying ≫
I remove the nasties lurking in your room

Safety ⚠
I'm safe around your furry friends and little people

Size ↗
Up to 1.8 m (6 ft)

JO'S TOP TIP

To keep your plant looking full and symmetrical, rotate it a quarter turn on a weekly basis.

Cast Iron Plant

ASPIDISTRA ELATIOR

The Cast Iron Plant is a real crowd-pleaser – with its gorgeous tall, dark green, paddle-shaped leaves, it makes a statement in any room. Super easy to look after, this plant is well known for being able to withstand some neglect, hence its name. It works well in dim spots and in places where other plants may have failed to thrive.

This plant has also been called the Bar Room Plant, because it would still be happy in the smokiest and dingiest of bars. While you can place it anywhere, try and avoid direct sunlight so its leaves don't get scorched.

There really isn't too much to tell you when it comes to care advice, as this plant will still thrive in the hands of the most forgetful plant parent and the darkest of corners in your home. However, you can show it love by wiping its leaves with a damp cloth to remove dust and unblock its pores so it can photosynthesise and grow, and repot every 4–5 years.

HOW TO KEEP ME HAPPY

Light ☀
I like bright indirect light

Water ◊
Water me when my top 5 cm (2 in) of soil feel dry to the touch

Air-purifying ≫
I remove the nasties lurking in your room

Safety ⚠
I'm safe around your furry friends and little people

Size ↗
Up to 90 cm (3 ft)

JO'S TOP TIP
The Cast Iron Plant is a slow grower, so don't be disheartened if yours produces 1–2 new leaves per year – this is normal.

Hen and Chicks Succulent

ECHEVERIA

Echeveria are perfect plants for beginners as they're incredibly easy to care for and can still thrive even if you neglect them a little. They grow in a pretty rosette shape and are native to Mexico and central and southern America.

This desert succulent from the Crassulaceae family is also known as Hen and Chicks. This is because Echeveria succulents grow quickly and produce offsets (clone-like daughters) called 'chicks'. Cute, hey!

Care is very simple for these succulents – give them bright, indirect light and don't overwater. Less is more! Always allow the soil to dry out between watering and it will appreciate a regular dusting of its leaves.

HOW TO KEEP ME HAPPY

Light ☼
I like bright indirect light

Water ◌
Let me dry out between watering

Wellbeing ♡
I can help you to relax and boost your mood

Safety ⚠
I'm safe around your furry friends and little people

Size ↗
Up to 31 cm (12 in)

JO'S TOP TIP

Don't worry if the outer leaves start to look a little unhappy – it's normal for the leaves closest to the soil to eventually shrivel and drop, and is the natural cycle of a succulent.

Ivy

HEDERA HELIX

You may be used to seeing this timeless climbing shrub outdoors, but Ivy is just as happy inside, too. It's easy to care for and quick growing, so is ideal to drape from a high shelf or hanging basket.

The word 'Helix' in this plant's name translates from Greek to mean 'twist and turn', which is a reference to how the vines rotate as they grow.

To keep your Ivy happy and growing well, it needs a bright sunny spot, so position it somewhere where a lot of light is guaranteed. It will appreciate regular misting to keep its leaves looking their best and remember to only water when its soil has dried out. If you're unsure, try the finger test on page 121.

HOW TO KEEP ME HAPPY

Light ☼
I like bright, indirect light

Water ◊
Water me when my top 5 cm (2 in) of soil feel dry to the touch

Air-purifying ≈
I remove the nasties lurking in your room.

Safety ⚠
Keep me out of reach of furry friends and little people

Size ↗
Up to 1 m (3 ft)

JO'S TOP TIP

Your Ivy may start to look leggy or spindly if it is reaching for a light source – if this happens, trim back and move closer to the light.

ELLE DECORATION N°289 SEPTEMBER 2016 NEW SEASON SIMPLICITY ELLEDECO
ELLE DECORATION N°342 FEBRUARY 2021 WELCOME TO THE FUTURE ELLEDECO
ELLE DECORATION N°335 JULY 2020 THE HEAT IS ON ELLEDECO
ELLE DECORATION N°339 NOVEMBER 2020 THE ARCHITECTURE ISSUE ELLEDECO
ELLE DECORATION N°326 OCTOBER 2019 DECORATE WITH PASSION ELLEDECO
ELLE DECORATION N°330 FEBRUARY 2020 INTRODUCING 2020 ELLEDECO
ELLE DECORATION N°329 JANUARY 2020 GET THE PARTY STARTED ELLEDECO
HOLLYWOOD ON'T HORSE THROUGH-SELF-IN LESIB 3HL THE BEST IN INTERNA
HOUSE & GARDEN APRIL 2021 EASY LIVING ELLEDECO
ELLE DECORATION N°275 JULY 2015 THE JOY OF COLOUR ELLEDECO
ELLE DECORATION N°312 AUGUST 2018 HAPPY DESIGN ELLEDECO
ELLE DECORATION N°338 DECEMBER 2019 FESTIVE FLAIR ELLEDECO
ELLE DECORATION N°299 JULY 2017 THE WELLBEING ISSUE ELLEDECO
ELLE DECORATION N°349 SEPTEMBER 2021 TIME FOR CALM ELLEDECO
ELLE DECORATION N°232 APRIL 2020 THE COMFORT ZONE ELLEDECO
ELLE DECORATION N°235 NOVEMBER 2013 COMFORT & JOY ELLEDECO
ELLE DECORATION N°274 JUNE 2015 THE NEW SIMPLE ELLEDECO
ELLE DECORATION N°340 DECEMBER 2020 SPRING GREENS ELLEDECO
ELLE DECORATION N°310 JUNE 2018 MAY 2019 ELLEDECO
DEFINING STYLE FOR MORE THAN 100 YEARS ELLEDECO
THE ART SPECIAL ELLEDECO
DEFINING STYLE FOR MORE THAN 100 YEARS ELLEDECO
AUGUST 2021 ELLEDECO
JHE CROSS-REVOLUTION 020 S/S FORECAST JHE
MARCH 1993 ELLEDECO
HdEL OCTOBER 2021 ELLEDECO

Mistletoe Cactus

RHIPSALIS BACCIFERA 'OASIS'

This is one cool-looking plant – with zigzagging fronds it will certainly make a statement in your home. Native to central and southern America, it is epiphytic, meaning it would be found naturally growing on and around trees in the rainforest.

Its Latin name is derived from the Ancient Greek word for 'wickerwork', referencing the plant's form. To ensure it thrives, it needs to be well drained – use a plant pot with a saucer and line the saucer with a few pebbles so it's never sitting in water. It likes being warm and humid, so is a great bathroom plant.

It can be sensitive to overwatering, too (the roots can rot if kept wet), so just water when the top layer of soil feels dry to the touch. This plant will let you know when it is thirsty because the leaves will lose their rigidity.

HOW TO KEEP ME HAPPY

Light ☼
I like bright indirect light

Water ◊
Water me when my top 5 cm (2 in) of soil feel dry to the touch

Wellbeing ♡
I can help you to relax and boost your mood

Safety ⚠
Keep me out of reach of furry friends and little people

Size ↗
Up to 55 cm (21 in)

JO'S TOP TIP
Just like us, this cactus needs regular haircuts! If its ends look dry, trim them to encourage healthy growth.

Whale Fin

SANSEVIERIA 'VICTORIA'

If you want a plant to stand out and make a statement, while still being minimalistic in style, then this is it! It may only have the one leaf – which resembles a whale's fin – but that is where its appeal lies.

This plant has one of the highest conversion rates for turning carbon dioxide into oxygen. Impressive! In the wild, it can grow up to 4 m (13 ft) tall. But it won't happen overnight – this plant is a slow grower.

The Whale Fin is very easy to care for and is tolerable to neglect. Just make sure you don't overwater it to avoid root rot.

HOW TO KEEP ME HAPPY

Light ☼
I like bright indirect light

Water ⬡
Water me when my top 5 cm (2 in) of soil feel dry to the touch

Air-purifying ≋
I remove the nasties lurking in your room

Safety ⚠
Keep me out of reach of furry friends and little people

Size ↗
Up to 90 cm (3 ft)

JO'S TOP TIP

This plant's leaf is its pride and joy, so wipe it down with a damp cloth to remove excess dust. Be careful not to use a leaf shine as it can be sensitive to chemicals

Areca Palm

DYPSIS LUTESCENS

For the holiday-at-home feel, look no further than the Areca Palm. This lush plant resembles a palm tree and its tropical vibes create your own little slice of paradise. It is also praised for its air-purifying qualities.

Native to the Madagascan tropics, this plant is sometimes referred to as the Golden Feather Palm or Butterfly Palm. It likes bright light, just keep it out of direct sunlight to avoid scorching its leaves.

If you can, use rain water when hydrating this plant, as it doesn't like the chemicals typically found in tap water. As for repotting, the Areca is a slow grower, so do this every 2–3 years – it likes to be cosy in a tight container.

HOW TO KEEP ME HAPPY

Light ☼
I like bright indirect light

Water ◊
I like my soil to be kept moist

Air-purifying ≋
I remove the nasties lurking in your room

Safety ⚠
I'm safe around your furry friends and little people

Size ↗
Up to 1.8 m (6 ft)

JO'S TOP TIP
If you notice the leaves turning brown, this could be an indicator of cold draughts or the air being too dry. Try moving to a warmer spot or give it a mist to add some humidity.

Sago Palm

CYCAS REVOLUTA

Native to southern Japan, the Sago Palm is a beautiful ornamental plant with bright green feathery leaves and spiky tips that spread out around a central cone. Despite its common name and deceptive appearance, the Sago Palm isn't a true palm tree. It is actually part of the Cycads family that has been around since prehistoric times – it has even been called a 'living fossil'.

Don't worry about the Sago Palm taking over your living room, as it is a slow-growing plant and will take around five years to reach its maximum height. Rotate it regularly to ensure even growth.

The Sago Palm is easy to care for, but will require frequent misting to replicate its natural tropical environment. Dust occasionally to clear the pores and help the plant breathe. Keep your palm out of direct sunlight too, as it will burn its leaves. Aim for indirect light.

HOW TO KEEP ME HAPPY

Light ☼
I like bright indirect light

Water ◊
Water me when my top 5 cm (2 in) of soil feel dry to the touch

Air-purifying ≋
I remove the nasties lurking in your room

Safety ⚠
Keep me out of reach of furry friends and little people

Size ↗
Up to 80 cm (2 ft 7 in)

JO'S TOP TIP

Wherever you choose to position this plant in your home, make sure there is room for its wide fronds to stretch out. In warmer months, you can even position it outside.

Sedum Burro's Tail

SEDUM MORGANIANUM

The Sedum Burro's Tail is one cool-looking plant, with its alien-like, distinctive trailing stems covered with plump green leaves. It's native to southern Mexico and is a real sun-worshipper – in fact its superpower is that it grows stronger in bright light. You may even see flowers during the summer, with pink or red star-shaped blooms emerging at the tips of the stems.

This plant is occasionally referred to as Donkey Tail Sedum as the fleshy stems dangle down just like, yep you guessed it, a donkey's tail!

This plant is a succulent and stores water in its leaves. During summer months, only water when its soil feels dry to the touch. It will need less frequent watering during the winter months.

HOW TO KEEP ME HAPPY

Light ☼
I like bright indirect light

Water ◊
Water me when my top 5 cm (2 in) of soil feel dry to the touch

Wellbeing ♡
I can help you to relax and boost your mood

Safety ⚠
I'm safe around your furry friends and little people

Size ↗
Up to 90 cm (3 ft)

JO'S TOP TIP

Handle with care! The stems of this plant are delicate and its leaves will break off if handled or moved, but new plants can be easily propagated from these fallen leaves.

African Milk Tree

EUPHORBIA TRIGONA CACTUS

This unique-looking plant may have cactus in its name but it's actually a succulent. It is native to West Africa but also found in tropical Asia and India. It has spines along its stems and small leaves that cover the edges of the stems and stalk. It loves direct sunlight, so is perfect for sunny windows.

This plant's common name, African Milk Tree, is a reference to the milky sap contained in the stems. It goes by many other names, including candelabra cactus, cathedral cactus, friendship cactus and good luck cactus – take your pick! It grows quickly, too, which is part of the fun.

The African Milk Tree is easy to care for, but remember that it will need dusting to unblock its pores, helping it to breathe. It also has a tendency to be top-heavy due to its shallow roots and heavy stems, so support the plant with canes when needed.

HOW TO KEEP ME HAPPY

Light ☼
I like bright direct light

Water ◊
Let me dry out between watering

Wellbeing ♡
I can help you to relax and boost your mood

Safety ⚠
Keep me out of reach of furry friends and little people

Size ↗
Up to 1.8 m (6 ft)

JO'S TOP TIP

Wear gloves when handling and pruning this prickly plant as its sap is toxic and can irritate skin.

Spider Plant

CHLOROPHYTUM COMOSUM

A classic, well-known houseplant, the Spider Plant can be found naturally in the South Pacific and South Africa. It has distinctive variegated foliage which resembles large blades of grass and, with the right care, the Spider Plant will produce small flowers that eventually turn into Spider Plant babies or Spiderettes!

There are over 200 species of Spider Plant and in the wild, they can grow to around 60 cm (2 ft).

This is a very low-maintenance houseplant, ideal for new plant parents or those with busy lifestyles. Its leaf tips can go brown and this is quite normal, just mist regularly and trim as required. Don't worry – it's very fast-growing so it will bounce back in no time!

HOW TO KEEP ME HAPPY

Light ☼
I like bright indirect light

Water ◌
Water me when my top 5 cm (2 in) of soil feel dry to the touch

Air-purifying ≋
I remove the nasties lurking in your room

Safety ⚠
I'm safe around your furry friends and little people

Size ↗
Up to 60 cm (2 ft)

JO'S TOP TIP

Stuck for gift ideas? Try propagating one of its babies in a jar of water – 3 weeks later it will be ready to plant – the perfect present for a loved one!

Bunny Ear Cactus

OPUNTIA MICRODASYS

The pads of this cactus can resemble bunny ears and its hair-like spines (glochids) look like fur, hence its affectionate common name. The Bunny Ear Cactus is native to northern Mexico and Arizona's desert regions, and is one of the most popular unkillables around.

In the wild and when it matures, the Bunny Ear Cactus can spread and grow outwards to cover 60–150 cm (2–5 ft) of ground – although it won't get quite as big as this in your home.

Contrary to what new cactus-keepers think, cacti require regular watering, at least during the summer. Only water the Bunny Ear Cactus again when the soil has dried out. Over the cooler months it may need only light watering every 3–4 weeks.

Always wear gloves when handling this plant to avoid skin coming into contact with the hairs of this cactus.

HOW TO KEEP ME HAPPY

Light ☼
I like bright indirect light

Water ⬦
Water me when my top
5 cm (2 in) of soil feel dry
to the touch

Wellbeing ♡
I can help you to relax and
boost your mood

Safety ⚠
Keep me out of reach of furry
friends and little people

Size ↗
Up to 60 cm (2 ft)

JO'S TOP TIP

You might not think it, but it's fun and easy to propagate – simply remove one of its ears and place into soil to create some baby bunnies!

Croton Plant

CODIAEUM VARIEGATUM

The striking Croton Plant never fails to get people talking! A natural show-off, it is native to Asia and the Western Pacific region. Most people fall in love with its variegated hues, interesting shapes and glossy autumnal-coloured leaves.

The Croton can be a little bit sensitive to change – when you repot, return it to the same position in your home.

This plant enjoys consistently moist (but not soggy!) soil and a regular misting. A tell-tale sign of a thirsty Croton Plant is drooping leaves. This plant is a sun worshipper, so position in your brightest southeast-facing room – it can even cope with direct sunlight.

HOW TO KEEP ME HAPPY

Light ☼
I like bright indirect light and can even cope with occasional direct sunlight

Water ◊
I like my soil to be kept moist

Air-purifying ≋
I remove the nasties lurking in your room

Safety ⚠
Keep me out of reach of furry friends and little people

Size ↗
Up to 1 m (3 ft)

JO'S TOP TIP

These leathery leaves require regular dusting to keep their pores unblocked and to encourage healthy growth.

Ponytail Palm

BEAUCARNEA RECURVATA

The ponytail palm is native to Mexico, so is tolerant of dry and barren landscapes, with the ability to hold water within its trunk. Its long and floppy leaves that grow from the top of the trunk resemble a ponytail.

Whilst its leaves are described perfectly by its common name, this plant is also known as Elephant's Foot – the perfect description for its chunky trunk!

This is a very forgiving plant, and can even go 2–3 weeks without water – not that I am suggesting you purposely neglect your Ponytail Palm! Remember, it has a natural dormant period in winter so will not need as much, if any, watering and feeding.

HOW TO KEEP ME HAPPY

Light ☼
I like bright indirect light and can even cope with occasional direct sunlight

Water ◊
Let me dry out between watering

Wellbeing ♡
I can help you to relax and boost your mood

Safety ⚠
I'm safe around your furry friends and little people

Size ↗
Up to 1.8 m (6 ft)

JO'S TOP TIP

As the Ponytail Palm can tolerate direct sunlight, it will naturally grow towards light, so rotate it a quarter turn every month to ensure it grows evenly.

Brazilian Cactus

PILOSOCEREUS CHRYSOSTELE

This Brazilian cactus is famed for its blue skin and fine hairs. They are tree-like in appearance and can grow tube-shaped flowers.

In the wild, these can grow up to 3 m (10 ft) tall. They're total sun worshippers – the brighter the spot, the better!

These are fast-growing cacti that need good drainage to ensure they are not sat in water. Being native to Brazil, they like warm and tropical conditions. Provided you protect them from the cold and avoid overwatering, these plants are very easy to care for.

HOW TO KEEP ME HAPPY

Light ☼
I like bright indirect light

Water ◌
Water me when my top 5 cm (2 in) of soil feel dry to the touch

Wellbeing ♡
I can help you to relax and boost your mood

Safety ⚠
I'm safe around your furry friends and little people

Size ↗
Up to 3 m (10 ft)

JO'S TOP TIP

When it comes to watering, I recommend the ice cube trick – pop one on the soil once a week and it will slowly melt, providing the perfect amount of water.

Steudner's Dragon Tree

DRACAENA STEUDNERI

This variety of Dragon Tree originated in the tropics of Africa, where it is often used as a decorative outdoor plant. It is fantastic for air purifying, making it ideal for the bedroom or office.

This plant is part of the asparagus family and gets its name from the Greek 'dracaena' (meaning 'female dragon') because the red resin in its stems that is said to resemble dragon blood.

This Dragon Tree will appreciate its leaves being wiped with a damp cloth to remove dust and unblock its pores, helping it to breathe and grow. It also likes regular misting, as this replicates the humid environment it is native to. To encourage a fuller plant, it can be pruned during growing season (April to September).

HOW TO KEEP ME HAPPY

Light ☼
I like bright indirect light, but am ok in shadier spots

Safety ⚠
Keep me out of reach of furry friends and little people

Water ◌
Water me when my top 5 cm (2 in) of soil feel dry to the touch

Size ↗
Up to 1.8 m (6 ft)

Air-purifying ≋
I remove the nasties lurking in your room

JO'S TOP TIP

It's all in the leaves with this plant – they may discolour if overwatered, and if underwatered they may appear to curl at the edges.

Golden Barrel Cactus

ECHINOCACTUS GRUSONII

The Golden Barrel Cactus is native to Mexico, where it has actually become quite endangered in the wild. This striking-looking cactus is globe-shaped, and as the name suggests, is perfectly round when young. It has deep ribs that bear long cream-coloured spikes that grow with age. It's sometimes called Mother-In-Law's Cushion, which I refuse to comment further on!

The Golden Barrel Cactus has a very slow growth rate, and although it can reach 50 cm (20 in) in height, this can take up to 20 years!

Like most cacti, this plant is very low maintenance and can cope with full and direct sunlight. During the warmer months, water when the top 5 cm (2 in) of soil has dried out. During the winter months, reduce the watering significantly and remember the golden rule: overwatering will cause its roots to rot, so less is more.

HOW TO KEEP ME HAPPY

Light ☼
I like bright indirect light

Water ⬭
Let me dry out between watering

Wellbeing ♡
I can help you to relax and boost your mood

Safety ⚠
I'm safe around your furry friends and little people

Size ↗
Up to 50 cm (20 in)

JO'S TOP TIP
Use an old paint brush to remove dust and cobwebs – it's tough going with a duster!

Dumb Cane

DIEFFENBACHIA CAMILLA

The Dumb Cane is found naturally growing in the jungles of Brazil. It's a very low-maintenance plant, recognised for its tall central stem with gorgeous oval, variegated leaves in shades of cream, yellow and green.

This plant's common name of Dumb Cane is due to the fact its leaves contain a toxic substance that can temporarily numb the vocal cords if consumed, but don't let that put you off – simply keep out of reach of paws and little ones.

Do not overwater this plant. Its soil should be moist not soggy, and reduce the amount of water given during the winter months. An occasional misting will also help to keep it looking its best. Add in a monthly dusting of its leaves to unblock pores and regular rotation to ensure even growth and you will have one happy plant!

HOW TO KEEP ME HAPPY

Light ☼
I like bright indirect light

Water ○
Water me when my top 5 cm (2 in) of soil feel dry to the touch

Air-purifying ≫
I remove the nasties lurking in your room

Safety ⚠
Keep me out of reach of furry friends and little people

Size ↗
Up to 1.5 m (5 ft)

JO'S TOP TIP
Occasionally the lower leaves turn yellow. This is perfectly normal but feel free to remove them so your plant doesn't waste its energy on them.

Zebra Cactus

HAWORTHIA

This cute and compact cactus is characterised by its chunky, rosette-shaped leaf clusters and white flecks. Think it looks familiar? Some compare Haworthias to Aloe Vera plants, which are all related and part of the Asphodeloideae family.

This plant is named after British botanist Adrian Hardy Haworth. It's very simple to propagate if you fancy growing your Zebra brood (page 28).

Native to South Africa, this plant has a very simplistic and easy care routine. You can get away with watering it as little as once a month, or simply when the soil has dried out. Just avoid watering the crown or rosette of the plant as this can upset it; instead pour straight into the soil. It can also cope with occasional direct sunlight.

HOW TO KEEP ME HAPPY

Light ☼
I like bright indirect light

Water ◌
Let me dry out between watering

Wellbeing ♡
I can help you to relax and boost your mood

Safety ⚠
I'm safe around your furry friends and little people

Size ↗
Up to 20 cm (8 in)

JO'S TOP TIP

This cactus may be a slow grower, but don't be disheartened – if you treat it well it can produce flowers in the height of summer.

Peace Lily

SPATHIPHYLLUM

Stunning in looks and serene in spirit, the Peace Lily is a great all-rounder. You'd be forgiven for thinking it may be difficult to care for, judging by its beautiful appearance; however, it is very low maintenance and, according to NASA, one of the top air purifiers.

According to the Chinese art of Feng Shui, Peace Lilies can cleanse the energy of a room, bringing peace and tranquillity to your home. If you give it plenty of light, you may be rewarded with rapid growth. If its leaves start to droop or sag, give it a drink and you'll see it perk up in no time.

The Peace Lily is happiest in bathrooms, where the humidity feels like its jungle home.

HOW TO KEEP ME HAPPY

Light ☼
I like bright indirect light

Water ⬨
Water me when my top 5 cm (2 in) of soil feel dry to the touch

Air-purifying ≋
I remove the nasties lurking in your room

Safety ⚠
Keep me out of reach of furry friends and little people

Size ↗
Up to 60 cm (2 ft)

JO'S TOP TIP

Don't panic if your Peace Lily's leaves or flowers wilt – simply remove them to encourage new growth.

Jade Necklace

CRASSULA MARNERIANA

A succulent, but with a twist – with its tightly stacked leaves and rose-coloured edges, this plant gets its common name from its cool and unusual appearance. It's native to South Africa and Mozambique and will make for a real eye-catching and unique addition to your green family. It's sometimes called a Worm Plant, which doesn't do its beauty justice!

There's no such thing as the winter blues with this plant, as it can produce small star-shaped flowers during the colder months. It looks fabulous trailing from shelves or hanging from a planter.

The Crassula is very easy to care for as it holds its water in its stems, making it ideal for beginners. It requires very little watering, and always allow the soil to dry out before watering again.

HOW TO KEEP ME HAPPY

Light ☼
I like bright indirect light, and can even cope with occasional direct sunlight

Water ◊
Let me dry out between watering

Wellbeing ♡
I can help you to relax and boost your mood

Safety ⚠
Keep me out of reach of furry friends and little people

Size ↗
Up to 20 cm (8 in)

JO'S TOP TIP

As a general rule, the Crassula is tolerant of crowded conditions, so can live happily in the same container for many years.

Velvet Calathea

CALATHEA RUFIBARBA

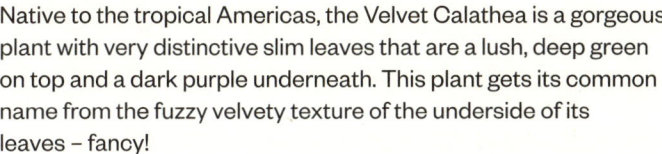

Native to the tropical Americas, the Velvet Calathea is a gorgeous plant with very distinctive slim leaves that are a lush, deep green on top and a dark purple underneath. This plant gets its common name from the fuzzy velvety texture of the underside of its leaves – fancy!

Not only does this plant look special, it is also great at purifying the air, making it perfect for your office or bedroom. At maturity, the Velvet Calathea can reach up to 1 m (3 ft) in height, so if it's getting too big for your desk, it'll look fab on the floor.

The Velvet Calathea enjoys a humid environment, so regular misting is important. Occasionally dust or wipe leaves with a damp cloth to help keep pores clear. Aim to keep soil moist but not soggy, and try not to allow the soil to dry out completely. As with most unkillables, reduce the amount of water given during the winter months.

HOW TO KEEP ME HAPPY

Light ☼
I like bright indirect light

Water ◊
I like my soil to be kept moist

Air-purifying ≋
I remove the nasties lurking in your room

Safety ⚠
I'm safe around your furry friends and little people

Size ↗
Up to 1 m (3 ft)

JO'S TOP TIP

If you notice the leaves on this plant turning brown at the edges, increase the amount of humidity it gets. An easy way to do this is to group your tropical plants together to create their own microclimate

Heartleaf Philodendron / Sweetheart Plant

PHILODENDRON SCANDENS

Be prepared to fall in love with this plant for so many reasons! For one, its common names of Heartleaf and Sweetheart Plant are a reference to the romantic heart-shaped leaves. Secondly, its botanical name 'philo' can be translated to mean 'loving' in Greek, and 'dendron' means 'tree', so it is literally a loving tree.

There are 489 different species of Philodendron – some serious plant goals to own them all! This is a fast-growing plant, so feel free to give it a trim if the vines get too long. Don't be worried if the new leaves look a little pale – they will darken with age.

This is a hardy and easy-to-care-for houseplant. It will benefit from an occasional dusting to keep the pores open. Ensure it's kept moist with weekly watering and regular mistings; however, the soil must be completely dry before topping up.

HOW TO KEEP ME HAPPY

Light ☀
I like bright indirect light

Water ◌
Water me when my top 5 cm (2 in) of soil feel dry to the touch

Air-purifying ≫
I remove the nasties lurking in your room

Safety ⚠
Keep me out of reach of furry friends and little people

Size ↗
Up to 1.5 m (5 ft)

JO'S TOP TIP

Vines can be pinched below a leaf node to encourage fuller growth – use the removed stem to have a go at propagating.

Dragon Tree

DRACAENA MARGINATA

Perfect for new plant parents, and praised for being one of the easiest indoor trees to grow, this plant can survive in the darkest corners of your room. We'd recommend you move it centre stage though, as it's quite the spectacle!

While the Dragon Plant is tall, it's a slow-grower, so don't expect it to change a lot. Enjoy it for its great air-purifying qualities, removing everyday toxins found in the air.

Show the Dragon Tree love with humidity and frequent misting of its leaves. Keep its soil moist, and only top up when it starts to dry out.

HOW TO KEEP ME HAPPY

Light ☼
I like bright indirect light

Water ◊
Water me when my top 5 cm (2 in) of soil feel dry to the touch

Air-purifying ≋
I remove the nasties lurking in your room

Safety ⚠
Keep me out of reach of furry friends and little people

Size ↗
Up to 1.8 m (6 ft)

JO'S TOP TIP

This houseplant is known to shed its bottom leaves as it grows, so try not to worry as this is usual behaviour. Simply cut them off so the plant can use its energy for new growth.

Birds Nest Fern

ASPLENIUM NIDUS CRISPY WAVE

Ferns have got a bit of a reputation as being divas in the plant world
– but don't let this put you off. The Birds Nest Fern is as hardy as
they come. A real keeper – not only does it have awesome wavy
leaves, it is also one of the longest-living pot plants around! It is
great for purifying your home's air, too.

The Crispy Wave is an epiphyte, so within its native habitat of
south east Asia it's common to find it growing on the branches
of other trees rather than in soil.

Rotate this plant regularly to ensure its growth is even. It's also
partial to frequent leaf misting, or pop it in the bathroom to ensure
it gets that humidity naturally from the steam caused by your
shower.

HOW TO KEEP ME HAPPY

Light ☼
I like bright indirect light

Water 💧
Water me when my top
5 cm (2 in) of soil feel dry
to the touch

Air-purifying ≫
I remove the nasties lurking
in your room

Safety ⚠
I'm safe around your furry
friends and little people

Size ↗
Up to 80 cm (2 ft 7 in)

JO'S TOP TIP

Keep your Crispy Wave at least 1 m (3 ft)
away from a heat source or draughty
windows, or its leaves, as the name
suggests, might indeed turn crispy!

PART 3

Plant SOS

FIRST AID

PLANT HACKS

GLOSSARY

First Aid

Overwatering

The phrase 'kill them with kindness' can be taken quite literally when applied to overwatering houseplants. Don't worry if you've been a little too overzealous – fix this by allowing excess water to drain away and give the soil time to dry out. Always do the finger test (see opposite) before watering. Also, move your plant to a shaded spot. When a plant is overwatered it struggles to transport water to its upper leaves. As a result, the top part of the plant can be in danger of drying out if left in direct sunlight.

Browning/crispy leaves

If you notice your plant's leaves are looking a little lacklustre, this is usually a common sign of low humidity which causes foliage to curl and go crispy. Remedy this by ensuring your plant isn't sat near a heat source like radiators or cold draughty windowsills, and mist it regularly with lukewarm water.

Dropping leaves

Shock is the most common cause of leaf drop. Plants are sensitive souls, so if you suddenly change the conditions they're used to (this can be anything from light to temperature) or if they're adjusting to a new home, they can experience stress. Don't panic, however, this is only temporary, and eventually, once your plant has adjusted to the new environment, it'll return to good health, but try to avoid moving your plants around too regularly.

Plant not growing

It can be frustrating to witness your plant doing very little. If you want to encourage growth, look to troubleshoot this by reassessing your plant's light and water needs. Getting too little or too much of one or both can be a cause of stunted growth. Consider the season – you can expect to see most growth in the summer months, and you can always add fertiliser to your plant during this period. Finally, another reason for little growth could be down to your plant becoming rootbound and needing to be repotted into a larger container (page 24).

Bugs

If you've noticed mini flies hovering around your plant, don't worry as this is completely normal and almost expected if you have a large green family. To get rid of them, make a soapy solution using washing-up liquid and warm water. Put this in a mister and spray your plants with it twice a week and the bugs should be gone in as little as 7 days.

Spindly growth

The dream is strong and healthy growth, but the reality is gangly and thin stems – sounds familiar? When plants start to look spindly in their appearance, this is due to them growing rapidly and weakly in search of a light source. You can correct this by researching how much light your plant requires and then move its location accordingly. Try trimming stems at the start of spring to encourage regrowth and help your plant grow back thicker and fuller.

Yellow leaves finger test

Leaves looking a little more yellow than green? There's a high chance incorrect watering is to blame! This could be either too much or too little. A good way to check if your plant needs water is to push your finger in the top 5 cm (2 in) of soil and dependent on how damp or dry it feels will dictate whether your greenery needs hydrating – this is also known as the finger test.

Roots spilling out

SOS, we've had a root outbreak! If your plant's roots are escaping the pot, it's definitely time to repot into a bigger container (page 24).

Sad/droopy plant

If your plant is looking a little sad and it's not clear as to why, have a look under the bonnet. Tap the nursery pot away from the bottom of the plant's soil and check the roots – if they're white and strong that means it's healthy, if they're black and soft it could be a sign of disease, overwatering or infestation.

Plant Hacks

Being a plant parent can sometimes be tricky, which is why it pays to have some genius plant tips in your pocket to help your green family thrive.

THE ICE CUBE HACK

Perfect for those who are prone to plant neglect or simply too busy living life to remember to water their plants on a regular basis. Position an ice cube on your plant's soil surface – just make sure you keep it away from the leaves. The ice will slowly melt, allowing the soil and plant to hydrate over time. Clever! Use this hack on plants that require little water – aim for one ice cube a week for small tabletop plants, two to three for larger ones.

THE CHOPSTICK HACK

The humble chopstick can actually help you decipher when your plant needs watering. Insert a chopstick into the soil – if it comes out with bits of soil attached to it, it is feeling hydrated. If it comes out bare, it's time to water! It's similar to inserting a knife into a cake when you're trying to figure out if it's finished baking.

THE SPONGE HACK

Your kitchen sponge can do more than just wash dishes! The next time you repot your plant, position a sponge in the bottom of the pot, beneath the soil. Here, it will absorb any excess water and works as a little reservoir, hydrating your plant's roots. Another great hack if you're always forgetting to water your thirsty plants!

BOTTOM-UP WATERING

When your plants are in need of watering (see the finger test on page 121) they can benefit from bottom-up watering. Place your pot in a saucer or tray filled with a couple of centimetres of water. Your plant should absorb this through the bottom of its soil for up to 30 minutes or until the soil feels moist to the touch. This keeps the roots evenly watered.

RAINWATER

Watering your plants with rainwater is a great way to improve their health. Rainwater is 'softer' than what comes out of your tap and is less likely to upset your plant. It also contains more oxygen, helping your plants to breathe better.

PEBBLE TRAY

To increase humidity, line a saucer with small stones or pebbles. Add water, but make sure that the pebbles are still poking up above the surface – this means your plant won't be sitting in water. Top up water as it runs dry – simple!

Glossary

AIR PURIFYING

Air-purifying plants are brilliant as they absorb harmful toxins from the atmosphere, making the air you breathe cleaner.

DORMANT

The plant equivalent of energy-saving mode; in other words, expect your green friend to slow down a bit. This usually occurs during the colder winter months, when they will show minimal activity or growth. They're perfectly healthy – just consider it hibernation.

EPIPHYTE

This term refers to a type of plant that grows on other plants, rather than in soil. You could see these plants attaching themselves on the branches of trees, for example. Usually, these plants will absorb water through their leaves or have modified roots. Plants really are clever beings!

FAMILY

A Family refers to a collection of plants that share botanical characteristics. They may have similar features such as a flower shape or appearance. To simplify how plant species are classified, it goes in order of genus, family, group. You'll know if a term refers to a plant family as it will end with the Latin suffix 'aceae' or 'ae', such as Araceae.

FROND

This is a type of leaf that is divided into sections – you'll likely see these on Fern or Palm plants.

GENUS

Plants are classified into a genus by features or characteristics.

HARDY

Another word for unkillable! Plants that are hardy are tolerant to difficult growing conditions, such as drought, shade or draughts.

HUMIDITY

Humidity is the relative amount of water vapour contained in the air. As temperature rises, the capacity of the air to hold water vapour increases, which makes bathrooms ideal for tropical plants. Replicate humid conditions for your tropical plants by using a mister or by grouping similar plants together to create a microclimate.

LEGGY

If your plant becomes spindly, develops long gaps between its leaves or loses them completely, it is what is known as leggy. Give it a prune (page 26) and move it closer to a light source to remedy.

PROPAGATION

This is an asexual means of reproduction – so by cutting the tip or stem of your existing plant, you can then produce a baby plant that is genetically identical to its parent (page 28). Plus, it saves you money and is very rewarding to watch a mini version grow.

PRUNE

To prune a plant, is to cut back and remove its leaves in order to promote healthy growth (page 26). We'd encourage you to do this as and when your plant needs it to allow it to retain a shape – and not to take over your home either!

REPOT

Depending on its growth rate, a plant will need to be repotted around every 1–2 years (page 24). Your thriving plant may need a larger sized pot to flourish in, or repotting can also mean changing the soil or potting mix of its current home (which is absolutely normal if your plant is a slow-grower). After all, the new soil will provide it with fresh nutrients. Both are vital to ensure a healthy plant that has the best growing conditions.

Acknowledgments

A huge thanks to Phillipa for making sense of my plant ramblings – her patience and humour has been essential in creating this book (and she makes a pretty good hand model occasionally, too – see page 9!).

Thanks to Jemma and Kay for making our plants look epic on every photoshoot; blood, sweat, laughter and the occasional therapy session makes them quite a dream team to work with.

Thank you Kate for all your hard work in bringing this book to life – your ideas and support mean a lot.

Big shout out to Luke, my partner in work and life – there's nobody I'd rather plant parent with.

And most importantly, thanks to my mum for dragging me around all those garden shows as a child and trusting me with my first succulent – who knew I'd grow up being green fingered like you?! X

About Jo Lambell

Jo Lambell is the founder of Beards & Daisies, one of the UK's largest online houseplant retailers. After living on Columbia Road for a decade, watching people awkwardly carry giant rubber plants and fiddle-leaf figs past her living room and across town, she left her corporate career and made it her mission to make buying houseplants easy. After attending horticultural college, Jo began bringing houseplants to the masses and sharing her industry expertise with her customers and the media. She believes that there's a perfect plant out there for everyone, and that we all have at least one or two green fingers.

Published in 2022 by OH Editions
Part of Welbeck Publishing Group.
Based in London and Sydney.
www.welbeckpublishing.com

Design © 2022 OH Editions

Text © 2022 Jo Lambell
Photography © 2022 Jemma Watts

A CIP catalogue record for this book is available
from the British Library.

ISBN 978-1-91431-723-1

Publisher: Kate Pollard
Editor: Gillian Haslam
Designer: Evi-O.Studio | Kait Polkinghorne
Illustrators: Evi-O.Studio | Emi Chiba, Zoe Gojnich
& Kait Polkinghorne
Stylist: Kay Prestney
Production controller: Arlene Alexander

Printed and bound by RR Donnelly in China

MIX
Paper from
responsible sources
FSC® C144853

10 9 8 7 6 5 4 3 2 1